Fluid Power
Educational
Series

# Hydraulic Filters:
# Construction, Installation
# Locations, and Specifications

Joji Parambath

*Hydraulic Filters: Construction, Installation Locations, and Specifications*

Copyright © 2026 Joji Parambath

**All rights reserved**

**ISBN:** 9798653640971

https://jojibooks.com

**First Edition 2020**
**Revised Edition 2023**
**Revised Edition 2025**
**Revised Edition 2026**

**Disclaimer of Liability**

The contents of this book have been checked for accuracy. We cannot guarantee full agreement, as deviations cannot be entirely precluded. Only qualified personnel should be allowed to install and work on pneumatic and hydraulic equipment. Qualified persons are defined as persons who are authorized to commission, ground, and tag circuits, equipment, and systems following established safety practices and standards.

# Table of Contents

# PREFACE

Filters must be integral parts of hydraulic systems to ensure proper operation of pumps, valves, and actuators. As hydraulic systems are demanding, the prescribed cleanliness levels of their fluid media must be met under all operating conditions. For this reason, it is essential to understand the different types of hydraulic filters and their performance ratings.

This book presents the principles of filtration in hydraulic systems. These principles include the filter media materials, various filter designs, and the typical filter locations in hydraulic systems. Further, this book describes the filter element performance ratings, such as the beta ratio and efficiency, and the multi-pass test used to determine them.

The same author gives many other fluid power topics in other textbooks under the fluid power educational series. A list of all the books is given at the end. Also, please see the details at https://jojibooks.com

Enjoy reading the book.
Your feedback is most welcome.

JOJI Parambath

# Chapter 1 | Constructional Features of Hydraulic Filters

A modern hydraulic system is susceptible to many types of contamination in its fluid medium. Contaminants can cause premature wear of internal surfaces, promote leakage, and clog flow paths. Water in the system fluid can lead to corrosion and accelerated component wear. Therefore, an efficient filtration system should be integral to every hydraulic system to separate particulate matter and water from the system fluid. Filters are installed at appropriate locations in a hydraulic circuit to control contamination effectively. Constructional features, types, installation locations, and specification parameters of filters are presented in the following sections.

## Function of Filters

Filters are necessary devices for removing particulate contamination from hydraulic systems. When fluid flows through a filter, it traps contaminants while allowing the fluid to flow easily, reducing unscheduled maintenance and downtime. As the filter clogs, the differential pressure across it increases until it bursts or collapses.

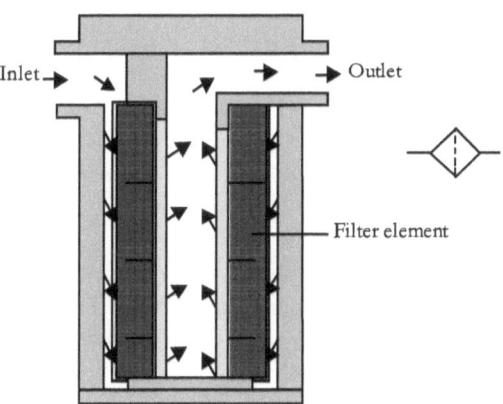

Figure 1.1 | The cross-section of a hydraulic filter

**Parts of Filters**

A hydraulic filter, as shown in Figure 1.1, consists of many essential parts and some optional parts. A filter consists of the following:

- Filter head
- Filter bowl
- Filter element

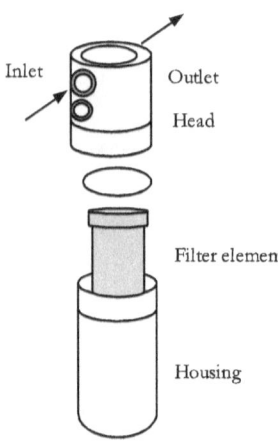

Figure 1.2 | An exploded view of a hydraulic filter

An exploded view of a filter is also shown in Figure 1.2. The filter element is housed in the filter bowl. The filter bowl and removable filter element (cartridge) can remain separate parts and be assembled onto the filter head as a cartridge-type filter unit. Alternatively, the filter element and the housing can be configured as an integrated unit, which can be screwed onto the filter head as a spin-on unit.

Hydraulic filters can be equipped with other useful features, such as bypass valves and clogging indicators. They can also be organized in duplex, in-line, or in-tank configurations to realize additional functions. Remember, various internal parts of a filter must be compatible with the fluid type. Let us now further elaborate on the filter parts.

## Filter Head

As shown in Figure 1.3, a filter head holds the filter element and its housing. It consists of ports for the inlet and outlet. It may include optional pressure gauge ports and visual and/or electrical indicators. It is made of cast Aluminium as a standard material or ductile iron for high-pressure applications.

## Filter Housing

A filter housing encloses the filter element. It also confines the system fluid within the unit. It must withstand the pressure within the unit. It is usually made of ductile iron or stainless steel. The hydraulic filter housing styles are cartridge, spin-on, duplex, in-tank, and in-line.

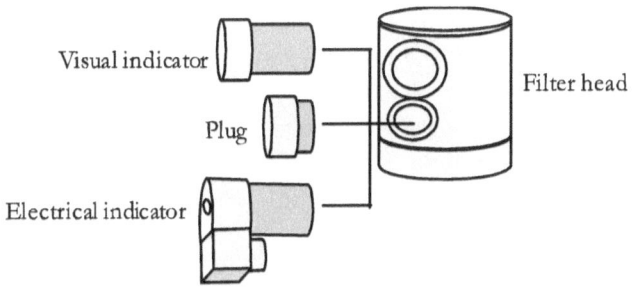

Figure 1.3 | Filter head

## Filter Element

A filter element (or cartridge) usually comprises a steel wire screen, cellulose, or synthetic glass fiber media. It consists of millions of tiny pores of micron sizes. A piece of the filter media is usually pleated and assembled into a canister as a disposable element (Figure 1.4). The pleats are made deeper and more numerous to provide the element with greater filtration surface area.

The fluid can flow from the outside of the filter element to the inside as in an 'out-to-in' filter or from the inside to the outside as in an 'in-to-out' filter.

Figure 1.4 | A filter cartridge

## Wire Mesh Filter Media

The wire-mesh media are made of epoxy-coated stainless steel (Figure 1.5). The filter captures contaminants in a fluid stream on one side of the wire screen facing the fluid flow. This type of filtration is regarded as surface filtration. A piece of wire-mesh filter media is used to make a coarse filter, usually called a strainer. Typically, a strainer traps very large, abrasive particulate matter that could tear a standard filter. Typical filtration ratings for wire mesh media include 25μ, 44μ, 50μ, 60μ, 74μ, 90μ, 100μ, 150μ, 200μ, etc.

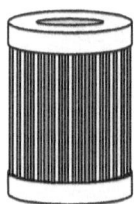

Figure 1.5 | A wire mesh filter cartridge

## Cellulose Filter Media

The cellulose (or paper) media, as shown in Figure 1.6, are made from plant fibers and are held together by resins. The pores in the cellulose media are microscopic. The multi-layered, thick-walled media absorbs contaminants throughout its depth as the fluid flows through it. Therefore, this type of media is known as depth media. The typical filtration ratings for cellulose media include 10 μm and 25 μm.

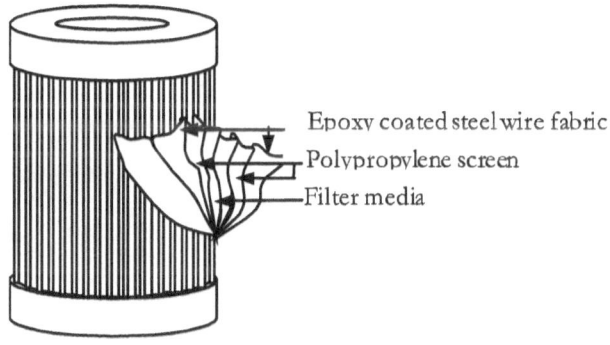

Figure 1.6 | A cellulose filter cartridge

## Glass Fibre Synthetic Filter Media

Synthetic glass fiber media are manufactured to be consistent and rounded to provide the least resistance to flow. It is made of inorganic micro-fine glass fibers. They are randomly laid into a multi-layered web with tapered pore geometry.

This type of construction ensures that larger pores are on the upstream surface and finer pores extend through the depth of the media. The thick-walled media captures contaminants throughout its depth. Therefore, a filter with synthetic media can produce high filtration efficiency.

Typical filtration ratings for glass fiber media include 3μ, 5μ, 6μ, 10μ, 20μ, 16μ, and 25μ.

## Surface Media Vs Depth Media

-**Surface Media** are made from woven wire. The filter captures contaminants in a fluid stream on one side of the wire screen facing the fluid flow.

-**Depth Media** are thick-walled filter media with absorbent materials. The filter usually absorbs contaminants throughout the depth of the filter media as fluid flows through it.

## Cartridge Type Filters

Initially, filters were cartridge type with replaceable cartridges in permanent housings, as shown in Figure 1.7(a).

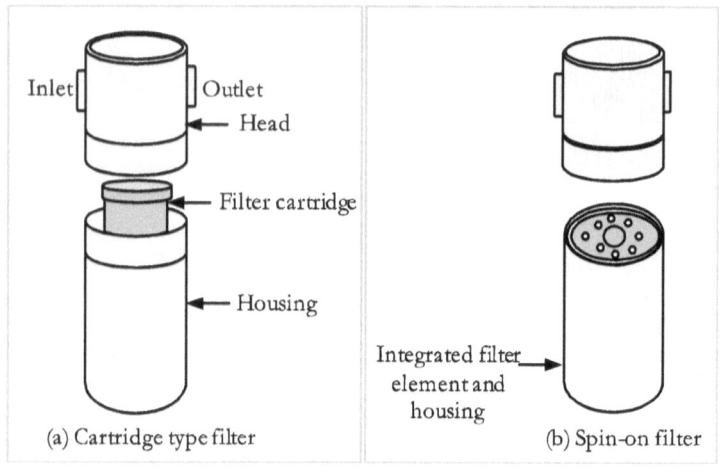

(a) Cartridge type filter       (b) Spin-on filter

Figure 1.7 | Hydraulic filters

## Spin-on Type Filters

As shown in Figure 1.7(b), this design is a self-contained filter element and housing assembly. The ineffective spin-on unit can be unscrewed from its mount and discarded. A new unit can then be screwed onto the mount. This design has made filter replacements more convenient. However, spin-on filters generated more waste with each filter unit replacement.

## In-tank Filter Type

In-line filters remove solid contaminants from hydraulic systems. They are space-saving units with simple screw-on covers and are ideal for low-pressure return-line applications. The filter head and ports, as shown in Figure 1.8(a), sit above the reservoir, while the housing lies within it. An in-tank filter is a heavy-duty unit with a die-cast aluminum head and a steel or nylon canister. An optional secondary inlet port provides a second return line.

## In-line Filter Type

An in-line filter removes solid contaminants from a hydraulic system, maintaining cleanliness and protecting components. As shown in Figure 1.8(b), the in-line filter assembly features a heavy-duty steel canister, a cartridge-type filter element, and an aluminum head and top cover. The steel housing design provides a strong, durable, and dependable unit. The filter cartridge can be replaced easily by removing the top cover.

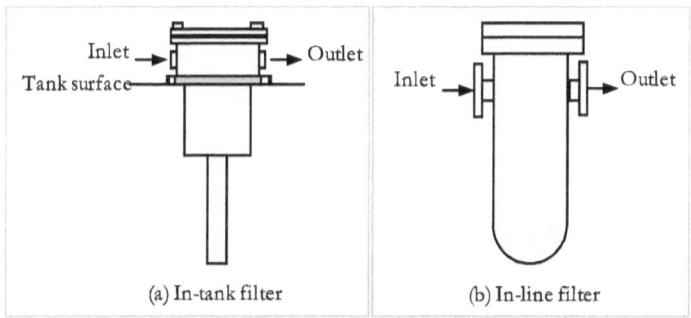

Figure 1.8 | Types of filters

## Filter with By-pass Valve

As shown in Figure 1.9, a bypass valve can be installed in a filter to protect against filter bursts. This filter assembly is a relief valve that diverts all or part of the fluid flow back to the system reservoir once the permissible maximum pressure differential across the filter is reached.

**Bypass Valve Setting:** Bypass valves have cracking pressures typically in the range of 0.1 bar (2 psid) to 7 bar (100 psid). The choice of cracking pressure for a filter depends on the filter's location. Suction and return-line filters have lower settings than pressure-line filters.

The collapse/burst pressure of the filter element should be at least 1.5 times the full flow pressure drop across the bypass valve and may range from 3.4 to 10 bar [50 to 150 psid].

7

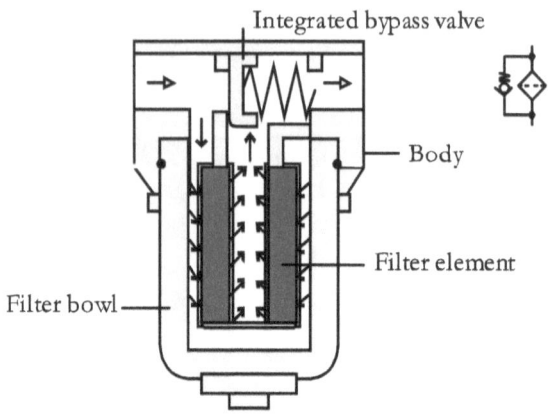

Figure 1.9 | A filter with a bypass valve

## Duplex Type Filter

It is an assembly of two (or more) filters with a selector valve, as shown in Figure 1.10, to select one filter at a time for active filtration. When the filter element in the selected filter becomes dirty and needs servicing, the valve is shifted without shutting down the system, thereby blocking flow through the ineffective filter and permitting flow through the second filter. The clogged filter element can then be cleaned or replaced while flow continues to pass through the currently active filter.

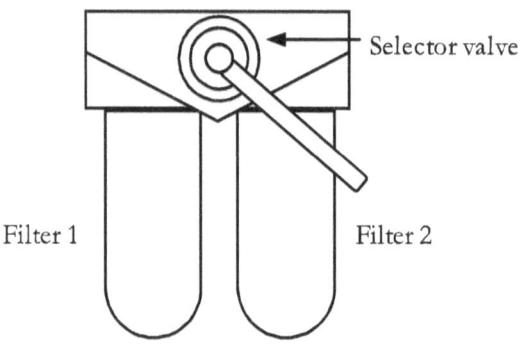

Figure 1.10 | A duplex-type filter

## Full-flow Filtration

In full-flow filtration, the entire fluid volume passes through the filter element during each cycle, eliminating the possibility of contaminated fluid reentering the system.

## Proportional-flow Filtration

In proportional or partial-flow filters, a portion of the fluid entering the filter passes through the filter element. At the same time, the remainder bypasses the filter element and flows through a venturi without filtration.

## Service Indicators / Contamination Indicators

A hydraulic filter element is efficient only if its particle-capture efficiency is fully exploited. For this purpose, a contamination indicator or clogging indicator is fitted to the filter. Vacuum indicators and switches can be used along with strainers to indicate when the vacuum reaches a certain level. A service indicator allows the filter element to be replaced only when its full dirt-holding capacity is reached, rather than replacing it on a time-based schedule. The contamination indicator is usually provided with a pressure gauge and a visual and/or electrical indicator to signal when its filter element needs to be replaced.

## Pressure Gauge

As shown in Figure 1.11(a), a pressure gauge measures pressure differential directly across the filter element. It consists of a mechanical bourdon tube with an elastic chamber, a link, a geared sector, a pinion, a pointer, and a calibrated scale.

Pressure applied to the elastic chamber causes the Bourdon tube to move outwards. The higher the pressure, the greater the tube's deflection radius. The deflection converts the applied pressure into the corresponding movement of the attached pointer across the scale via links, levers, and gearing.

The pressure gauge is a safety measure that monitors overpressure in the system and assists with troubleshooting.

## Visual Indicator

Figure 1.11(b) shows two visual filter clogging indicators that provide real-time, color-coded updates - green for good filters and red for clogged ones - indicating when filter element replacement is needed. In Figure 1.11(b)(i), an indicator with a pointer moves through green, reaching red when the filter is fully clogged. Figure 1.11(b)(ii) shows a pop-up indicator with a green sleeve and a movable red sleeve. When the red sleeve completely covers the green, it indicates full filter clogging.

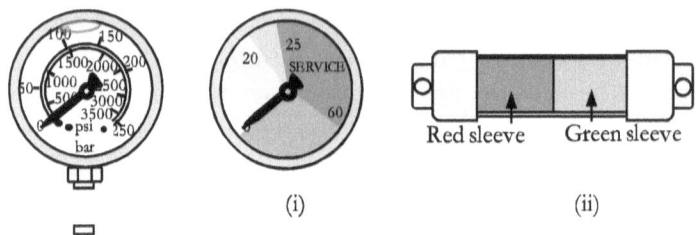

(i)                                     (ii)

(a) Pressure gauge            (b) Visual indicators

Figure 1.11 | Visual indicators for hydraulic filters

## Electrical Indicators

Engineers use electrical/electronic sensors when connections to remote sensing devices, such as alarms, horns, or lights, are desired. Electrical indicators are available in a 2-wire or 3-wire design. Symbols of different types of electrical indicators are shown in Figure 1.12.

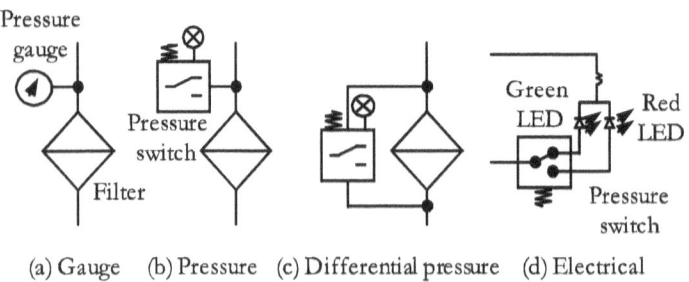

(a) Gauge     (b) Pressure   (c) Differential pressure   (d) Electrical
indicator        switch                switch                     part

Figure 1.12 | Symbols of service indicators

## Seals

Filters with Buna-N seals are appropriate for most applications involving petroleum-based oil. Filter seals made of fluorocarbon, such as FPM (FKM), EPDM, etc., are used for applications operating at temperatures above 83°C [150°F] and for fluids such as diester, phosphate ester, water-glycol, water/oil emulsions, and HWCF.

Note:
**FPM** is the international abbreviation according to DIN/ISO
**FKM** is the short form for the fluoroelastomer according to ASTM
**EPDM** (Ethylene Propylene Diene Methylene rubber)

## Magnet

The filter unit may include a magnetic column to attract and retain ferrous particles smaller than 1 μm. A filter with a magnet is shown in Figure 1.13.

Fluid flow through the filter element is typically from the outside to the inside (out-to-in). However, for effective filtration of high concentrations of metallic particles, a filter designed to direct flow from the inside of the filter element to the outside (in-to-out) can be used. The magnet removes ferrous particles from fluids that a standard filter cannot capture.

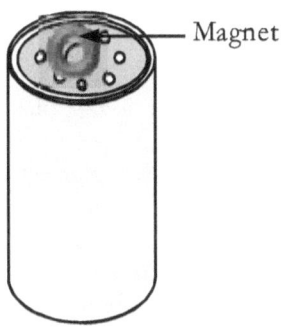

Figure 1.13 | A filter with a magnet

# Chapter 2 | Hydraulic Filter Locations

According to their locations in a hydraulic system, filters are categorized into four types: (1) Suction strainer/filter, (2) Pressure filter, (3) Return-line filter, and (4) Offline filter. An air filter is also used in a hydraulic system fitted on the top of the reservoir. Figure 2.1 shows the filter locations.

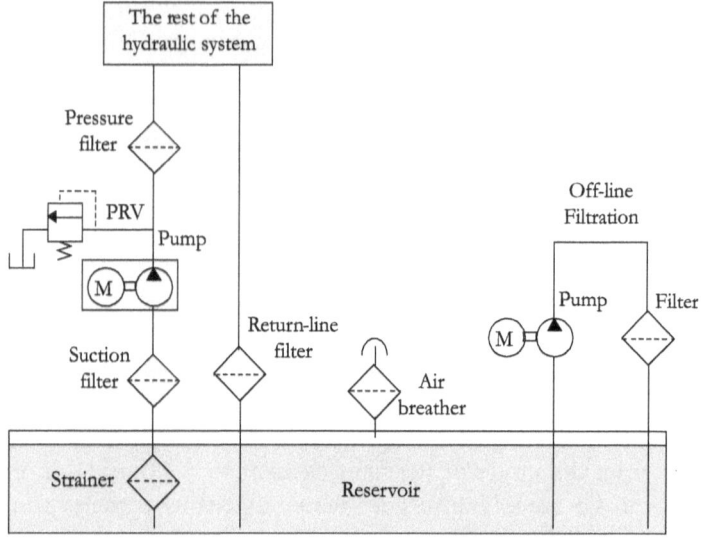

Figure 2.1 | Filter locations in a hydraulic system

**Strainer**
A strainer is a coarse filter with a zinc-plated housing, stainless steel wire-mesh screens, and a rugged steel core, usually with a mesh width of ≥149 microns. It is installed at the pump suction side. The main objective of a strainer is to protect the pump from coarse particles at the lowest possible cost. However, a blocked strainer can starve the pump, which, in turn, can cause cavitation. A strainer used in a hydraulic system is cleanable and reusable. However, it isn't easy to clean, as it is submerged in the system's reservoir. They offer the least benefit.

## Suction Filter

A conventional suction filter is mounted outside the reservoir, above the fluid level, in a service-friendly manner. The space-saving in-tank suction filter can be mounted semi-immersed at the fluid level.

A suction filter is a coarse filter with a mesh width typically 5 to 149 microns. It removes coarse particles from the fluid before entering the hydraulic pump. That is, it protects the pump from coarse particles at an economical cost. A bypass valve with low cracking pressure can be incorporated into a suction filter to prevent the starvation of the associated pump.

Suction filters are less expensive and easier to service than other types of filters. However, they offer only medium benefits.

## Pressure Filter

A pressure filter is installed downstream of the pump. It is usually subjected to the maximum system pressure. It is constructed with a rugged casing to withstand the maximum system and shock pressures. The filter mesh width can be finer (10–20 microns). It must be sized for the specific flow rate.

The primary function of the pressure filter is to keep the fluid clean, removing contamination generated internally by the pump. It protects expensive, dirt-sensitive downstream components. Therefore, pressure filters offer great benefits.

A pressure filter should have an internal bypass valve with a cracking pressure typically in the range from 2.5 to 8 bar [35 to 115 psi] to protect the filter from collapse/burst.

Pressure filters are available in medium-pressure up to 140 bar [2000 psi] and high-pressure up to 450 bar [6500 psi] ratings.

They are available in cartridge, spin-on, in-line, and/or duplex configurations.

However, they are more expensive since their housings are pressurized. Servicing a pressure filter is also tricky because of its heavy-duty construction.

## Return-line Filter

A return filter is installed in the return line. Remember, a return line typically sees pressure in the range of 20 to 30 bar [300 to 435 psi]. It is usually low-pressure housing. It can tolerate a higher pressure differential across its filter element. It removes particles ≥25 microns in mineral-based fluids and ≥10 microns in synthetic fluids. The purpose of the return-line filter is to trap dirt from the system's working components and particles entering the system through any worn seals on those components. Return-line filters are available in the in-tank, in-line, and duplex configurations.

Return-line filters offer high benefits. However, they may be subject to flow surges during operation. A relief valve may be placed across the filter to provide an additional path for the excess return flow.

## Off-line Filtration / Kidney-loop Filter

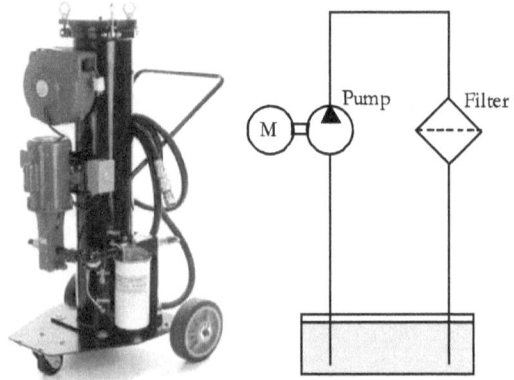

Figure 2.2 | An image of an offline filter unit
Courtesy: Engineered Filtration, Inc., USA

As shown in Figure 2.2, the offline filtration system consists primarily of a separate pump, filter unit, and hoses. The filter unit uses a fine filter element and a low-pressure housing that are easy to service. The components of the offline filtration system can be arranged on a mobile cart. An offline filtration system with quick-disconnect couplers can be temporarily retrofitted to an existing system. One offline filtration unit can serve many hydraulic systems. Alternatively, an offline filtration system can be permanently integrated into a system as a dedicated unit.

In the offline filtration system, fluid is continuously pumped from the reservoir, passed through the filter, and returned to the reservoir. It is a subsystem that operates independently of the primary hydraulic system. It is used to achieve the best possible filtration results in hydraulic systems. However, this type of filtration system requires an additional pump-motor unit.

**Air Breather**

As shown in Figure 2.3, an air breather unit typically consists of air filters and sufficient desiccant material. It is fitted to the reservoir in a hydraulic system to protect the fluid against airborne particulate contamination and moisture.

The filter removes solid particulates typically down to 3 microns, and the desiccant material prevents moisture from entering the reservoir in a hydraulic power pack.

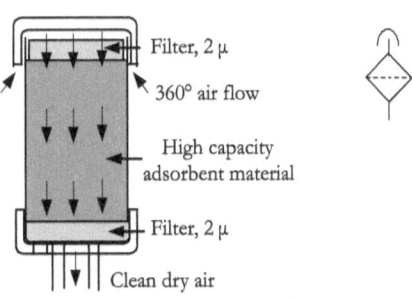

(a) Air breather

Figure 2.3 | A cross-sectional view of an air breather

# Chapter 3 | Performance Ratings of Filters

The degree of cleanliness achieved by the fluid in a hydraulic system can be linked to the performance of the filter elements used. These elements are rated based on their ability to separate contaminants of particular sizes from the system fluid under specific test conditions.

Filter manufacturers publish various performance data for their filters. The parameters specified by the filter manufacturers are the mesh number, Beta (ß) Ratio, filter efficiency, and/or the micron ratings. Besides, there is an industry standard called the 'multi-pass test' for effectively measuring filter performance ratings.

## Mesh Number/Sieve Number
The mesh size (or fineness) of a wire-mesh filter can be expressed as its mesh number (or sieve number). It is the number of openings from the center of any wire of the wire mesh to the center of a parallel wire one inch away, as illustrated in Figure 3.1.

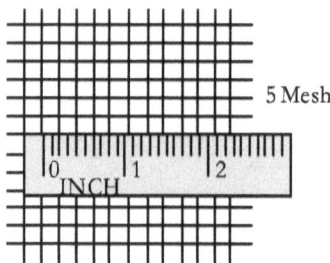

Figure 3.1 | Illustrating the measurement of mesh number

## Mesh Number to Micron Size Conversion
A chart for converting some important mesh numbers to microns is given in Table 3.1. Table A1.1 in Appendix 1 provides a detailed conversion chart from mesh to microns.

Table 3.1 | Mesh numbers to microns

| Mesh Number | Microns |
|:---:|:---:|
| 60 | 250 |
| 100 | 149 |
| 140 | 105 |
| 170 | 88 |
| 200 | 74 |
| 230 | 62 |
| 270 | 53 |
| 325 | 44 |
| 400 | 37 |
| 550 | 25 |
| 800 | 15 |
| 1250 | 10 |
| 2500 | 5 |

## Beta Ratio

It is a measure of a filter element's filtration efficiency. The test setup to measure the Beta ratio of a test filter is shown in Figure 3.2. The Beta ratio can be determined by monitoring contamination levels upstream and downstream of the filter during a single pass. That is,

$$\text{Beta ratio, } \beta_{(x)} = \frac{\text{Particle count in the upstream oil}}{\text{Particle count in the downstream oil}}$$

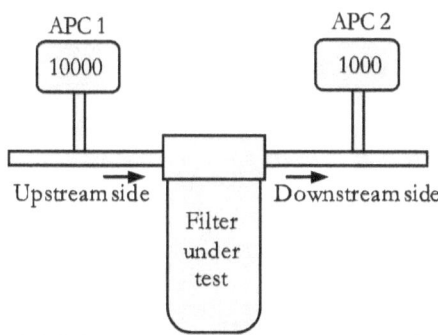

Figure 3.2 | A test setup for measuring the Beta ratio

## Filter Efficiency

Filter efficiency has the same meaning as the beta ratio. It is given by:

$$\text{Efficiency}_{(x)} = (1 - \frac{1}{\beta}) \cdot 100$$

Table 3.2 | Beta ratio and corresponding efficiencies

| Upstream particles ($\geq$x μm) | Downstream particles ($\geq$x μm) | Beta (ß) ratio | Efficiency $_{(x)}$ |
|---|---|---|---|
| 1,00,000 | 50,000 | 2 | 50.0% |
| 1,00,000 | 5,000 | 20 | 95.0% |
| 1,00,000 | 1,333 | 75 | 98.7% |
| 1,00,000 | 1,000 | 100 | 99.0% |
| 1,00,000 | 500 | 200 | 99.5% |
| 1,00,000 | 100 | 1000 | 99.9% |

Table 3.2 gives Beta ratios and the corresponding efficiencies.

## Example 3.1

**In a single-pass test for finding the performance of the test filter, 100000 particles of size $\geq$6 microns are counted upstream using a laser particle counter, and 1000 particles of the same size range are counted downstream of the filter. Calculate the Beta ratio and the filter's efficiency.**

Solution

Upstream particle count, $C_u$ = 100000
Downstream Particle count, $C_d$ = 1000

Beta ratio, $\beta_{(6)}$      = $C_u / C_d$
Beta ratio, $\beta_{(6)}$      = (100000 / 1000) = 100

Efficiency $_{(6)}$      = $(1 - \frac{1}{\beta}) \times 100$
     = $[1 - (1/100)] \times 100\%$
     = 99%

## Micron Ratings

The micron rating of a filter indicates its ability to remove a specified percentage of particles of a certain size.

- **The absolute micron rating** of a filter is the smallest size of particles it can capture, over 98.6% on the first pass through it.

- **The nominal micron rating** of a filter is the smallest micron size of particles it can capture in a specified quantity, ranging from 50% to 95% on the first pass through it.

For example, the filtration efficiencies of samples from filters manufactured in the same batch are measured at different micron sizes and recorded in Table 3.3. From the Table, the absolute micron rating is 5 microns, and the nominal micron rating is 2, 3, or 4 microns. It may be noted that paper filters are usually rated nominally, and synthetic filters are usually rated in absolute terms.

Table 3.3 | Sample test data for efficiencies against μ sizes

| Micron size | Filtration efficiency |
|-------------|----------------------|
| 20 μ | 99.9% |
| 10 μ | 99.1% |
| 5 μ | 98.61% |
| 4 μ | 90% |
| 2 μ | 50% |

## Differential Pressure (ΔP)

Differential pressure (ΔP) across the filter indicates the difference between its inlet and outlet pressures.

## Particle Capture Efficiency

It is the weight of the specified artificial contaminant (ISO medium test dust) that must be added to the fluid upstream of the filter to produce a given pressure differential across the filter (usually 1 bar) under the specified conditions. It indicates how much solid dirt the filter element can hold before it needs to be replaced.

## The Multipass Test

The multi-pass test gives reproducible test data for assessing the performance of hydraulic filter elements. This test provides an accurate, universally accepted method for describing the efficiency of a filter element in removing particles of standard test dust from the test fluid over a wide range of particle sizes under controlled laboratory conditions. This test is standardized by ISO 16889, SAE J1858, ANSI, and NFPA.

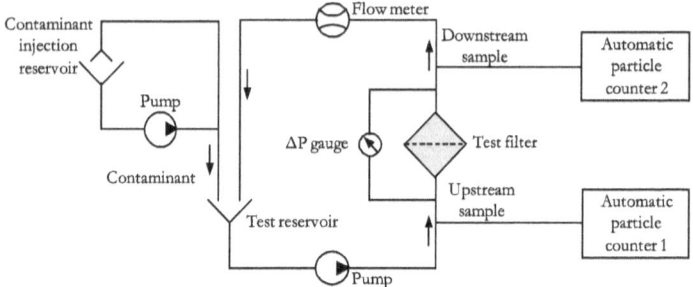

Figure 3.3 | A test setup for a multipass test

A setup for conducting the multipass test is shown in Figure 3.3.

## Filter Selection

Filters are essential for maintaining hydraulic systems at a satisfactory level of cleanliness. Therefore, their proper selection is essential to the hydraulic system design.

For this purpose, it is necessary to know the system characteristics, such as the flow rate, working pressure, fluid viscosity, and expected level of system contamination. It is also essential to consider the system's cleanliness level requirements.

The filters should meet their performance specifications and align with the system requirements to achieve the desired level of cleanliness. The performance specifications of each filter/filter element, such as its micron rating, Beta-ratio/filter efficiency, dirt-holding capacity, and the maximum permissible pressure differential across the filter, must be considered.

Other important factors, such as pressure rating, reliability, service life, and filter cost, are no less important.

The compatibility of the filter construction materials with the fluid used in the system must also be considered.

Summarizing the above points, consider the following factors when selecting the filter:

- Flow rate and working pressure

- Oil viscosity, temperature, and target fluid cleanness level

- Filter element media

- Micron ratings

- Beta ratio

- Filter efficiency

- Dirt holding capacity

- Maximum allowable pressure drop

- Seal options

-Fluid type

**Note:**
Recommended filtration levels for hydraulic fluids are given in Appendix 2.

*See Chapter 6 for details on filter sizing.*

# Chapter 4 | Typical Filter Specifications

Table 4.1 | Filter parameters

| Parameter | Choice (Typical) |
|---|---|
| Filter element media | - Stainless wire mesh (30 bar)<br>- Stainless fiber (210 bar)<br>- Cellulose<br>- Inorganic glass fiber (30 bar/210 bar) |
| Micron rating | 3, 5, 10, 20, 25, 50, 100, 200 µm |
| Flow rate | 60, 110, 160, 240, 330, 475, 660, 990 lpm<br>[14, 30, 45, 70, 90, 125, 160, 250 gpm] |
| Maximum operating pressure | Depends on the type of filters, namely suction filters, pressure filters, and return-line pressures |
| Seal material | - NBR<br>- FPM (Viton)<br>- EPDM (Ethylene Propelene Diene Monomer) |
| Valve | - Bypass valve<br>- Reverse flow valve<br>- Non-return valve |
| Bypass setting | 0.1 bar to 7 bar<br>[2 psid to 100 psid] |
| Clogging indicator | - Visual<br>- Electrical<br>- Pressure gauge |
| Port size | - NPT ¾ to 3<br>- G ½ to 2<br>- GAS and SAE Flanges<br>- SAE O-ring |

# Chapter 5 | Water Removal Media

Water is undesirable in mineral oils and synthetic hydraulic fluids. Dissolved and free water in hydraulic fluids can be removed using water-removal media, vacuum dehydration, or other methods.

The water removal media can be prepared with desiccants or a coalescer. Vacuum dehydrator units can be used when the water concentration level in hydraulic fluids exceeds 100 ppm (0.01%).

The following paragraphs explain some of the methods to remove water from hydraulic systems:

**Adsorption:** The media part is a highly adsorbent copolymer laminate with an affinity for water.
**Absorption:** A filter with a water-absorbing element can absorb free water in the fluid medium of a hydraulic system up to 100% saturation.
**Centrifuge**: This method separates free water from the fluid chamber using centrifugal force generated by rapidly rotating cone-shaped disks. This method can remove only up to 100% of free water.

**Coalescer:** A coalescer is designed to remove free water from a fluid by passing it through its unique coalescing elements. This method can also remove only up to 100% of free water.

**Vacuum Dehydrator:** In this method, free and dissolved water can be removed using a vacuum dehydration unit with a vacuum pump, a circulation heater, and dispersion media. The vacuum pump draws the fluid into the unit through the heater, where it is heated to approximately 65°C [150°F]. The fluid is then exposed to a vacuum while it flows through the pores of the dispersion media. The water is boiled off from the unit, and the fluid is effectively dehydrated. This method can remove water down to 20% saturation.

# Chapter 6 | Hydraulic Filter Sizing Factors

A filter consists of a filter element and a housing. It may include an inbuilt bypass valve biased with a spring, providing a parallel path around the filter element to protect it from collapsing when heavily loaded with contaminants.

Bypass valves have cracking pressures typically ranging from 1 bar (15 psid) to 7 bar (100 psid). Return-line filters have a lower setting than pressure-line filters.

The filter element should be replaced immediately when the differential pressure indicator signals the need for service.

A filter must be sized based on the maximum flow rate through the filter, the maximum system pressure, the permissible pressure drop across the filter, the operating fluid temperature, and the required filter element service interval. Let us elaborate on these points.

## Maximum Flow Rate
A filter should be sized based on its maximum flow rate. A system with multiple double-acting cylinders can produce a higher return flow rate through their blind ends when operated simultaneously. As a result, return-line filters can experience surge flows under such conditions.

## Maximum Pressure
Pressure-line filters are subjected to full system pressure, whereas return-line filters are subjected to lower pressure. A filter may experience pressure fluctuations, and in such cases, the fatigue rating of the filter housing must be considered.

## Permissible Pressure Drop
System pressure drops occur across a filter's housing and element. The maximum pressure drop occurs when the filter is fully clogged, and the entire flow goes over the bypass valve. For

example, a filter with a 2.5 bar (36 psid) bypass valve will have a pressure drop of 5 bar (70 psid) at the maximum flow through the bypass valve.

Excessive pressure drop across the filter may cause failure of upstream components, such as the hydraulic motor shaft seals. Select a filter with the worst-case pressure drop within limits.

### Temperature and Viscosity
During the cold startup of a hydraulic system, the fluid viscosity is often high enough to cause a very high pressure drop across the filter element. This will open the bypass valve briefly while the fluid warms up. In most cases, this condition is acceptable.

The fluid temperature also affects filter seals. Select seals that will withstand extreme temperatures without failure.

### Filter Element Service Interval
A filter element is replaced when the differential pressure across it indicates a need for replacement. The interval between filter element replacements depends on the permissible pressure drop, the beta ratio, and the filter's dirt-holding capacity.

A filter with the highest dirt-holding capacity can provide the longest service life, reducing the need for frequent element replacements and downtime.

### Fluid Type
Filters in a hydraulic system are generally sized for petroleum fluids. However, synthetic fluids are also used in hydraulic systems. Most filter manufacturers provide recommendations for sizing filters used in hydraulic systems with synthetic fluids.

### Filter Media Selection
The selection of filter media depends on the required cleanliness. Wire mesh and cellulose media elements are nominally rated, whereas synthetic filters are rated in absolute terms.

# Chapter 7 | Maintenance of Hydraulic Filters

Regular maintenance for filters and strainers in a hydraulic system includes checking for clogs and cleaning or replacing the media as needed. The reservoir must be drained completely to access the strainer installed in it.

The general process for cleaning a wire-mesh filter element begins by removing it and shaking the dust by hand. If the filter is heavily clogged, rinse it thoroughly with a neutral solvent, brush off debris and dirt, shake it to remove excess solvent, rinse it again with hydraulic fluid, and dry it in the shade.

A paper or synthetic filter element must be checked at least every three months. It should also be replaced as soon as its internal bypass valve opens or after a system repair, typically at least once every 12 months. Remember that impurities can be introduced into the filter during a system repair.

## Filter Service Steps, Typical

Hydraulic filters must be cleaned or replaced during routine system maintenance. However, most filters are not cleanable. A typical procedure for replacing a cartridge-type filter element is outlined below:

-Check whether the OEM-specified filter replacement interval has been reached
-Check the filter's clogging indicator
-Determine whether filter element replacement is needed
-Inspect the new filter for any damage

-Switch off the pump
-Relieve the system or trapped pressure completely
-Unscrew the housing by turning it counterclockwise; if necessary, use a wrench
-Remove the clogged filter element or cartridge
-Wipe the filter housing area with a clean cloth

-Remove any sediment present inside the housing
- Remove gaskets, O-ring, and housing seals
- Clean the filter head surfaces
- Check the filter's integrity
- Lubricate seals and threads with clean system fluid
- Install the new filter in the housing

- Fit the housing to the filter head per the instructions on the filter housing
- Do not over-tighten the housing to the filter head
- Check for leaks
- Properly dispose of the old filter element according to local regulations

## Useful Points, Hydraulic Filters
- Set up a filter maintenance schedule and follow it.
- Inspect filter elements removed from the system for signs of failure.

## Service Steps for Cartridge Filters
Figure 6.1 provides a quick reference for service steps involved with cartridge-type filters.

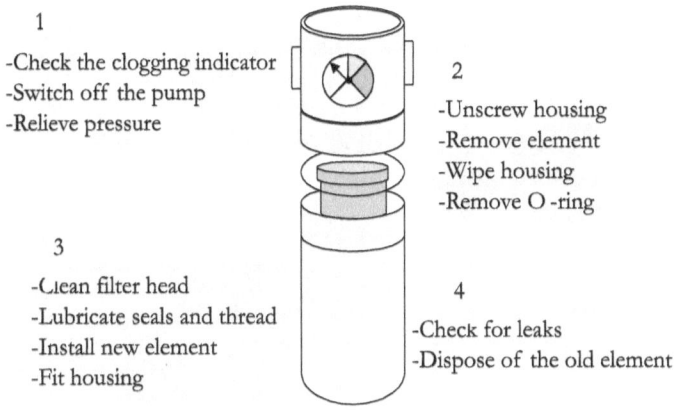

Figure 6.1 | Service steps for cartridge-type filters

# 8 | Objective Type Questions

1. Which type of hydraulic filter can be used to prevent the collapse of a filter due to clogging of its filter element?
a) Filter with bypass valve
b) Duplex filter
c) Proportional filter
d) Full-flow filter

2. A strainer employs:
a) Mechanical type, surface filtration
b) Mechanical type, depth filtration
c) Absorbent type, surface filtration
d) Absorbent type, depth filtration

3. Which type of hydraulic filter can you use for its easy maintenance without shutting down the process line?
a) Filter with bypass valve
b) Duplex type of filter
c) Proportional filter
d) Full-flow filter

4. A return-line filter generally employs:
a) Mechanical type, surface filtration
b) Mechanical type, depth filtration
c) Absorbent type, surface filtration
d) Absorbent type, depth filtration

5. Which type of hydraulic filter can you use to protect valves and cylinders in a hydraulic system immediately?
a) Suction filter
b) Pressure filter
c) Return-line filter
d) Offline filter

6. A strainer employs:
a) Cellulose media
b) Wire-mesh media
c) Synthetic media
d) Depth media

7. Which hydraulic filter can you use to filter fluid independent of the primary hydraulic system?
a) Suction filter
b) Pressure filter
c) Return-line filter
d) Offline filter

8. Synthetic filter media are made from:
a) Cellulose
b)  Glass fibers
c) Steel wires
d) Metal disks

9. Which type of hydraulic filter can be used to trap dirt from the working components of a hydraulic system?
a) Suction filter
b) Pressure filter
c) Return-line filter
d) Offline filter

10. Which of the following types of filters offers high benefit in hydraulic systems?
a) Strainer
b) Suction filter
c) Return-line filter
d) None

11. What is the efficiency of the hydraulic filter with a beta ratio of 20 for 4μ particles?
a) 50%
b) 80%
c) 95%
d) 99%

12. When a hydraulic filter performance is expressed as $ß_4 = 75$, what does it mean?
a) A beta ratio of 4, 75% efficient at removing 75-micron particles
b) A beta ratio of 75, 98% efficient at removing 4-micron particles
c) A beta ratio of 4, 75% efficient at removing 4-micron particles
d) A beta ratio of 75, 75% efficient at removing 4-micron particles

13. Which of the Beta Ratios given below provides the better wear protection at 10 microns?
a) $ß_5 = 75$
b) $ß_{10} = 200$
c) $ß_{25} = 200$
d) $ß_{200} = 10$

14. The pressure differential across a hydraulic filter unit increases due to:
a) The use of higher-viscosity fluid
b) The clogging of the filter media
c) The friction of the fast-moving fluid
d) All of the above

15. Which of the following accessories is not used to remove water from hydraulic systems?
a) Adsorption filter
b) Coalescing filter
c) Diffuser
d) Vacuum dehydrator

# 9 | Review Questions

1. Why is it necessary to install an efficient filtration system in a hydraulic system?

2. Explain the functional and constructional aspects of strainers used in hydraulic systems.

3. Classify hydraulic filters according to their nature of filtration.

4. Differentiate the surface and depth filter media used in hydraulic systems.

5. What materials are used to make the filter media for hydraulic filters? Briefly explain.

6. What are the advantages of synthetic filter media?

7. Differentiate the cellulose filter media and the synthetic filter media.

8. What is a bypass filter, as used in hydraulic systems?

9. Explain the working principle of a hydraulic filter with a bypass valve with the help of its schematic diagram.

10. Explain the function of a duplex-type hydraulic filter with the help of its schematic diagram.

11. What is a full-flow hydraulic filter?

12. What is a proportional flow hydraulic filter?

13. Explain how the bypass valve complements the full-flow filter when used in a hydraulic system.

14. Explain the significance of the location-based hydraulic filtration.

15. What are the typical locations where hydraulic systems can install filters?

16. Explain the essential function and installation features of suction filters in hydraulic systems.

17. Explain the essential function and requirements of pressure filters in hydraulic systems.

18. What are the disadvantages of suction filters?

19. Explain the primary function of the return-line filters used in hydraulic systems.

20. Enumerate the advantages and disadvantages of pressure filters in hydraulic systems.

21. What are the advantages of return-line filters?

22. Explain the hydraulic offline filtration technique with a neat circuit diagram.

23. What are the advantages of the offline filters used in hydraulic systems?

24. What are the performance specifications of hydraulic filters?

25. What is the difference between the absolute and nominal micron ratings?

26. What is the burst rating of a hydraulic filter?

27. Explain the multi-pass test for a hydraulic filter with the schematic diagram of the test setup.

28. What parameters can be determined through the multi-pass test for a hydraulic filter?

29. How is the efficiency of the hydraulic filter evaluated?

30. What is the Beta ratio of a hydraulic filter? Explain with a numerical example.

31. What is filter efficiency, and what is its significance in hydraulic systems?

32, How is the Beta ratio of a hydraulic filter related to its efficiency?

33. Explain the meaning of Beta ratio, $\beta_{4(c)} = 2$, of a hydraulic filter.

34. Write short notes on the following terms related to hydraulic filters: (1) Synthetic filter media, (2) Proportional flow filtration, (3) Multi-pass test, and (4) Dirt holding capacity.

35. Write short notes on the following terms of hydraulic filters: (1) Depth filter, (2) Duplex filter, (3) Offline filtration, and (4) Beta ratio.

36. What system characteristics do you consider when specifying a hydraulic filter?

37. What factors do you consider while selecting hydraulic filters?

38. What are the different ways to remove water from hydraulic systems?

# 10 | Numerical Problems

1. Find out the Beta ratio of the filter when 10000 particles of $\geq 20$ μm enter the filter, and 9950 of these particles are trapped by the filter. Also, calculate the filtration efficiency for the given particulate size.

[Ans: $\beta_{(20)} = 200$, $\eta_{(20)} = 99.5\%$]

2. What is the meaning of $\beta_4(c) = 1000$, of a hydraulic filter? Also, calculate the corresponding filter efficiency.

3. In a multi-pass test, 30000 and 1000 particles of size 4 microns or greater were recorded in automatic particle counters upstream and downstream of the test filter, respectively. Determine the filter's efficiency for a given particulate size.

[Ans: $\eta_{(4)} = 96.67\%$]

4. Filtration efficiencies corresponding to different micron sizes are obtained from single-pass tests of sample filters of the same batch. What is the absolute micron size of the batch of filters?

Table 9.1 | Sample test data for efficiencies against μ sizes

| Micron size | Filtration efficiency |
|:---:|:---:|
| 20 μ | 99.9% |
| 10 μ | 99.5% |
| 5 μ | 99.1% |
| 4 μ | 98.62% |
| 2 μ | 90% |

[Ans: Absolute micron size = 4 μ]

***Objective-type questions - answer key:***
*1-a, 2-a, 3-b, 4-d, 5-b, 6-b, 7-d, 8-b, 9-c, 10-c,*
*11-c, 12-b, 13-b, 14-d, 15-c*

# Appendix 1

## Mesh to Micron Conversion

Table A1.1: Mesh to Micron conversion chart

| U. S. Mesh | Microns | Millimeter | Inches |
|------------|---------|------------|--------|
| 3 | 6730 | 6.730 | 0.2650 |
| 4 | 4760 | 4.760 | 0.1870 |
| 5 | 4000 | 4.000 | 0.1570 |
| 6 | 3360 | 3.360 | 0.1320 |
| 7 | 2830 | 2.830 | 0.1110 |
| 8 | 2380 | 2.380 | 0.0937 |
| 10 | 2000 | 2.000 | 0.0787 |
| 12 | 1680 | 1.680 | 0.0661 |
| 14 | 1410 | 1.410 | 0.0555 |
| 16 | 1190 | 1.190 | 0.0469 |
| 18 | 1000 | 1.000 | 0.0394 |
| 20 | 841 | 0.841 | 0.0331 |
| 25 | 707 | 0.707 | 0.0280 |
| 30 | 595 | 0.595 | 0.0232 |
| 35 | 500 | 0.500 | 0.0197 |
| 40 | 400 | 0.400 | 0.0165 |
| 45 | 354 | 0.354 | 0.0138 |
| 50 | 297 | 0.297 | 0.0117 |
| 60 | 250 | 0.250 | 0.0098 |
| 70 | 210 | 0.210 | 0.0083 |
| 80 | 177 | 0.177 | 0.0070 |
| 100 | 149 | 0.149 | 0.0059 |
| 120 | 125 | 0.125 | 0.0049 |
| 140 | 105 | 0.105 | 0.0041 |
| 170 | 88 | 0.088 | 0.0035 |
| 200 | 74 | 0.074 | 0.0029 |
| 230 | 63 | 0.063 | 0.0024 |
| 270 | 53 | 0.053 | 0.0021 |
| 325 | 44 | 0.044 | 0.0017 |
| 400 | 37 | 0.037 | 0.0015 |

# Appendix 2

**Recommended Filtration for Hydraulic Fluids**

| ISO Code | NAS Code | Absolute Filtration (or better) |
|:---:|:---:|:---:|
| 14/12/9 | 3 | $\beta_{5\copyright} \geq 100\ (\geq 99\%)$ |
| 15/13/10 | 4 | $\beta_{5\copyright} \geq 100\ (\geq 99\%)$ |
| 16/14/11 | 5 | $\beta_{7\copyright} \geq 100$ |
| 17/15/12 | 6 | $\beta_{7\copyright} \geq 100$ |
| 18/16/13 | 7 | $\beta_{10\copyright} \geq 100$ |
| 19/17/14 | 8 | $\beta_{15\copyright} \geq 100$ |
| 20/18/15 | 9 | $\beta_{20\copyright} \geq 100$ |

For more details on the ISO and NAS codes, please refer to the textbook on 'Hydraulic Fluids' (Book No. 10) under the Fluid Power educational series by the same author. [See Page 33]

# 11 | References

1. Article on: 'Condition Monitoring for Hydraulic and Lubricating Fluids' HYDAC International GmbH, Hauptstrasse, Saarbrücken, www.hydac.com

2. Article on: 'Element Technical Data Fundamentals', SCHROEDER INDUSTRIES LLC

3. Article on: 'Filtration' Stauff Corporation, Inc., 7 Wm. Demarest Place, Waldwick, New Jersey, USA

4. Article on: 'Hydraulic & Lubrication Filters, Part I: Filter Types and Locations and Part II: Proper Filter Sizing, HY-PRO Filtration', www.filterelement.com

5. Article on: 'Hydraulic filtration - part 1, 2, 3, 4', Penton Media, Inc. & Hydraulics & Pneumatics magazine

6. Article on: 'Hydraulic Filtration and Contamination', Filter Manufacturers Council, NC, USA, www.filtercouncil.org

7. Article on: 'Hydraulic System Filters New ISO Fluid and Cleanliness Rating Standards', Moog Inc., NY, USA, 8. Articles on (1) 'The Micron Rating for Media in Fluid Filters', Revised October 2005, and (2) 'Hydraulic Filter Performance Criteria', Filter Manufacturers Council, Research Triangle Park, NC, USA

9. Document on 'MAHLE Industrial Filters, Filtration processes for the metalworking industry', MAHLE Filtersysteme GmbH, Industriefilter, Schleifbachweg, Öhringen, www.mahle.com

10. Document on: 'Filter Element Beta Ratio Information', Swift Filters, Inc., Ohio, USA

11. Document on: 'Hydraulic & Lubrication Filters, Part I: Filter Types and Locations', HY-PRO FILTRATION, U. S. A., www.filterelement.com

12. Document on: 'The Handbook of Hydraulic Filtration', Parker Hannifin Corporation, Metamora, OH, USA.

13. Documents on: 'Hydraulic Filtration Technical Reference' (Doc. No. F115354 rev.1), and 'Hydraulic Filtration Product Guide', Donaldson Company, Inc., Minneapolis, MN, USA.

14. Document on 'Hydraulic & Lubrication Filters, Part II: Proper Filter Sizing', WPFS 1005, Hy-Pro Engineering Department, HY-PRO Filtration

# Fluid Power Educational Series Books

1. Pneumatic Systems and Circuits -Basic Level (In the SI Units)
2. Industrial Pneumatics -Basic Level (In the English Units)
3. Pneumatic Systems and Circuits -Advanced Level
4. Electro-Pneumatics and Automation
5. Design of Pneumatic Systems (In the SI Units)
6. Design Concepts in Pneumatic Systems (In the English Units)
7. Maintenance, Troubleshooting, and Safety in Pneumatic Systems
8. Industrial Hydraulic Systems and Circuits -Basic Level (In the SI Units)
9. Industrial Hydraulics -Basic Level (In the English Units)
10. Hydraulic Fluids
11. Hydraulic Filters: Construction, Installation Locations, and Specifications
12. Hydraulic Power Packs (In the SI Units)
13. Power Packs in Hydraulic Systems (In the English Units)
14. Hydraulic Cylinders (In the SI Units)
15. Hydraulic Linear Actuators (In the English Units)
16. Hydraulic Motors (In the SI Units)
17. Hydraulic Rotary Actuators (In the English Units)
18. Hydraulic Accumulators and Circuits (In the SI Units)
19. Accumulators in Hydraulic Systems (In the English Units)
20. Hydraulic Pipes, Tubes, and Hoses (In the SI Units)
21. Pipes, Tubes, and Hoses in Hydraulic Systems (In the English Units)
22. Design of Industrial Hydraulic Systems (In the SI Units)
23. Design Concepts in Industrial Hydraulic Systems (In the English Units)
24. Maintenance, Troubleshooting, and Safety in Hydraulic Systems
25. Hydrostatic Transmissions (HSTs) (In the SI Units)
26. Concepts of Hydrostatic Transmissions (In the English Units)
27. Load Sensing Hydraulic Systems (In the SI Units)

28. Concepts of Load Sensing Hydraulic Systems (In the English Units)
29. Electro-hydraulic Proportional Valves
30. Electro-hydraulic Servo Valves
31. Cartridge Valves
32. Electro-hydraulic Systems and Relay Circuits
33. Practical Book: Pneumatics - Basic Level
34. Practical Book: Electro-pneumatics - Basic Level
35. Practical Book: Industrial Hydraulics – Basic Level
36. Programmable Logic Controllers and Programming Concepts
37. Compressed Air Dryers
38. Hydraulic Circuits – Identification of Components and Analysis

For more details, please visit https://jojibooks.com

# About the Author

Joji Parambath has been a trainer in Pneumatics, Hydraulics, and PLC for over 25 years. During his career, he has trained numerous professionals from the industry, faculty members, and engineering students.

He is the key trainer at Fluidsys Training Centre, Bangalore, India (https://fluidsys.org), which provides training in Pneumatics and Hydraulics. He has already written two books on Pneumatics and Hydraulics. The publication of the present series of 32 books is intended to restructure and update the existing books.

The author wishes to thank all trainees for their lively interaction and many useful suggestions during the training programmes that prompted the author to write the present series of books. You may send your feedback to joji.p@hotmail.com

10th June 2020

www.ingramcontent.com/pod-product-compliance
Lightning Source LLC
Chambersburg PA
CBHW031503210526
45463CB00003B/1060